BARNES & NOBLE® READER'S COMPANION™

A Brief History of Time

BARNES & NOBLE® READER'S COMPANION™

Today's take on tomorrow's classics.

FICTION

THE CORRECTIONS by Jonathan Franzen
I KNOW WHY THE CAGED BIRD SINGS by Maya Angelou
THE JOY LUCK CLUB by Amy Tan
THE LOVELY BONES by Alice Sebold
THE POISONWOOD BIBLE by Barbara Kingsolver
THE RED TENT by Anita Diamant
WE WERE THE MULVANEYS by Joyce Carol Oates
WHITE TEETH by Zadie Smith

NONFICTION

THE ART OF WAR by Sun Tzu
A BRIEF HISTORY OF TIME by Stephen Hawking
GUNS, GERMS, AND STEEL by Jared Diamond
JOHN ADAMS by David McCullough

BARNES & NOBLE® READER'S COMPANION™

STEPHEN HAWKING'S

A Brief History of Time

BARNES
& NOBLE
BOOKS

EDITORIAL DIRECTOR Justin Kestler
EXECUTIVE EDITOR Ben Florman
DIRECTOR OF TECHNOLOGY Tammy Hepps

SERIES EDITOR John Crowther
MANAGING EDITOR Vincent Janoski

WRITER Howard Rich
EDITOR Matt Blanchard
DESIGN Dan O. Williams, Matt Daniels

This edition published by Spark Publishing

Spark Publishing
A Division of SparkNotes LLC
120 Fifth Avenue, 8th Floor
New York, NY 10011

ISBN 1-58663-862-9

Library of Congress Cataloging-in-Publication Data available upon request

Printed and bound in the United States

Contents

BARNES & NOBLE® READER'S COMPANION™

WITH INTELLIGENT CONVERSATION AND ENGAGING commentary from a variety of perspectives, BARNES & NOBLE READER'S COMPANIONS are the perfect complement to today's most widely read and discussed books.

○ ○ ○

Whether you're reading on your own or as part of a book club, BARNES & NOBLE READER'S COMPANIONS provide insights and perspectives on today's most interesting reads: What are other people saying about this book? What's the author trying to tell me?

○ ○ ○

Pick up the BARNES & NOBLE READER'S COMPANION to learn more about what you're reading. From the big picture down to the details, you'll get today's take on tomorrow's classics.

Barnes & Noble® Reader's Companion™

A Brief History of Time

A TOUR OF THE BOOK

Life, the Universe, and Everything

A Brief History of Time addresses some of the most mind-boggling questions of the universe on an amazingly personal level.

WHEN PEOPLE SEARCH FOR THE ANSWERS to humanity's biggest questions, they usually consult one of two sources: their scripture or Stephen Hawking's *A Brief History of Time.* Many consult both. At a time when the word of science is the gospel truth, Hawking's book is one of the most vivid examples of how far science has indeed come. His book is unique in that it combines a rigorous scientific approach with the kind of childlike curiosity that first impels us to ask big questions—Where do we come from? What does the future hold for us? And of course, What is quantum chromodynamics? With the exception of this last one, Hawking doesn't pose these questions on an individual level, nor even on the level of mankind. He asks questions on the largest possible scale—the universal. In doing so, he approaches one of humankind's greatest questions: "Why do we exist?" And Hawking comes as close as anyone ever has to answering this question. Maybe this is why *A Brief History of Time* is, after the Bible, the most widely read book of our time.

Hawking has two goals in his book: to explain the basic principles of cosmology (the study of the natural order of the universe) and to discuss its future. Accordingly, the book has a two-part structure, although Hawking doesn't divide it explicitly. In roughly the first half of the *A Brief History of*

Time, he carefully traces the relevant history of cosmology up to our present understanding. He maps a direct course from Aristotle to Einstein, clearly explaining a connect-the-dots version of the history of physics and astronomy. In the book's second and more difficult part, Hawking connects the final dot: he draws a line from Einstein's revolutionary theories to his own. And with the updated, tenth-anniversary edition of *A Brief History of Time*, Hawking makes this connection stronger. In the years between editions, several scientific discoveries have proved a number of Hawking's hypotheses, and the scientific community has widely endorsed his findings.

What makes Hawking's book so engaging isn't just his remarkable proficiency as a scientist and teacher, but his own personality. Hawking is nothing like what we might expect from a brilliant scientist. Far from being professorial or aloof, he allows a vulnerable side of his character to creep in through the margins of his book. Occasionally, he infuses even the most abstract scientific discussion with personal drama. Describing the debate over the implications of Einstein's theory of general relativity, Hawking recalls his own days as a doctoral candidate. Recounting his struggle to make sense of Einstein's theory, he muses on his own struggles with ALS, or Lou Gehrig's disease (he was diagnosed with the disease at roughly the same time he tackled Einstein). *A Brief History of Time* is primarily a book of scientific theory, but it takes on a surprisingly personal dimension—something unexpected in a science book.

Hawking analyzes the universe with the fervor of a man trying to understand the mind of God.

Hawking's passionate quest to find meaning in the cosmos is a personal one, for his search for meaning in his own life becomes entwined with his search for meaning in the universe. Although Hawking dedicates his life to solving the larger questions of physics, he agonizes over whether he should continue his research despite the likelihood that he has no more than a year left to live. He doesn't say it outright, but his solution to this problem is soon obvious. Hawking's efforts to solve the great mysteries of the universe become an inquiry into the mystery of his own fate. He analyzes the universe with the fervor of a man trying to understand the mind of God.

General readers are attracted to *A Brief History of Time* because Hawking's voice is not entirely scientific, but also humble and vulnerable—in short, human. The famous cover of the book, which shows a picture of Hawking sitting in his wheelchair against a background of constellations, illustrates this idea perfectly. It is a poignant image, juxtaposing frail humanity against the elegant grandeur of the heavens.

But the image isn't a pessimistic one. *A Brief History of Time* makes us believe in a not-too-distant future in which we will have solved the great riddle of the cosmos. Hawking predicts that one day, we will perceive the grand design of the universe. The closer he gets to making this discovery, the more his findings encourage his belief in a divine order. Hawking's exploration of the cosmos is more than just science—it's a quest fueled by a desire to understand our place in the vastness of the universe.

CHAPTERS 1–3
A BRIEF HISTORY OF SCIENCE

Hawking's personal appeal is apparent from the first sentence of his book, which begins not with a scientific postulate but with a funny, accessible anecdote. In these introductory chapters, Hawking tries to devote as much time to discussing the people behind history's most momentous scientific theories as the theories themselves. He takes special delight in relating silly stories about misguided scientists whose theories seem quaint in retrospect.

Talking about the concept of a cosmic microwave background, Hawking tells us that its two discoverers, Arno Penzias and Robert Wilson, originally mistook it for bird feces lodged in their high-tech microwave detector. While introducing the concept of universal gravitation, Hawking vehemently denies the myth that Newton needed an apple to fall on his head to inspire his breakthrough. Einstein contemplated the theory of relativity while working as a clerk in the Swiss patent office. But above all, Hawking is interested in which scientists have received the Nobel Prize for their efforts and which have missed out (most conspicuously, Hawking himself).

Ultimately, science beats out the scientists in the struggle for Hawking's full attention—or rather, all the scientists aside from himself. Early in the book, Hawking defines his basic goal and the goal of all cosmologists: to formulate a single unified theory of the universe. It's interesting that the field of cosmology, which constantly grapples with concepts like infinite time and infinite space, has one single goal, one endpoint. The goal of cosmology is to formulate a "theory of everything"—a single theory that unifies all other theories in physical science. Applying such a theory to any event occurring in the universe, we would be able to explain not only how it happened but what will happen next.

A theory of everything once seemed as elusive as a theory of the elementary particles or a theory of the beginning of the universe, but now it seems close to being achieved. In fact, *The Theory of Everything* is even the provocative title of Hawking's newest book. We're close—after all, physical science can now be reduced to two theories that together can

> *"The eventual goal of science is to provide a single theory that describes the whole universe."*

explain almost everything: the theory of relativity and the theory of quantum mechanics. The problem is that the two theories are incompatible with each other where they meet. Relativity accurately explains large-scale physical phenomena, and quantum mechanics explains things on the tiniest atomic scale, but the two theories can't be used in tandem with any accuracy.

Though Hawking hasn't yet formulated a unifying theory, he does have a strategy for how he'll do it. He has found places in the universe—black holes—where the two incompatible theories literally collide. In the early parts of his book, Hawking doesn't explain how this collision happens. But he does establish a pattern of scientific discovery that serves as a template for his later innovations.

The first section of *A Brief History of Time* does more than simply summarize the history of science up to today. It also shows the method of thought that gives rise to new scientific theories. When reading Hawking's brief history of major scientific advances, we begin to notice a pattern: the truly revolutionary upheavals of science—such as Aristotle's assertion that the Earth is round, or Einstein's theory of general relativity—only came about through these thinkers' ability to reimagine our fundamental conception of the universe. By telling us stories about how skeptics dismissed these scientists in their own times, Hawking prepares us for his own revolutionary assertions. Indeed, some of Hawking's later points do seem ridiculous. If your first response to the notion of "imaginary time" isn't hysterical laughter, it's probably disbelief or, at the very least, mild annoyance.

Technical observations haven't yet proved all of Hawking's theories definitively. But in style, his theories resemble those of his most lofty predecessors. Hawking's ideas have solved contemporary problems in cosmology by redefining our entire conception of the universe. We can see that Hawking's own technique reflects those of the scientists he admires most, like Aristotle, Copernicus, and Einstein. Hawking is too tactful to make this parallel himself. Nonetheless, by understanding the ways in which past scientists have overcome popular misconceptions, we note a resemblance to Hawking's own iconoclastic way of thinking about science. Like his predecessors, Hawking has come up with a new understanding of the universe, not by trying to sew up problems within existing theories, but by enlarging our conception of the universe. His greatest achievements are those that have challenged the most trusted scientific truths of our time—and that show that these theories aren't completely accurate.

We can see a prototype for Hawking's strategy as early as Aristotle. In 340 B.C., Aristotle was troubled by a seeming contradiction: the popularly accepted notion was that the Earth was flat, but the North Star seemed to move along the horizon depending on the place on land from which it was viewed. This wandering tendency of the North Star seemed to indicate that it was some kind of exception to the rule that stars did not normally move. But instead of trying to figure out why the North Star might be a special case, Aristotle courageously chose to modify his entire view of Earth to accommodate this irregularity. He imagined a universe in which Earth wasn't, in fact, flat. He imagined his world in terms more complicated than had yet been conceived.

Copernicus found himself confronted with a similarly difficult problem. How could he reconcile Earth's position at the center of the universe with the fact that the other planets didn't seem to make regular orbits around Earth? Rather than search for ways to explain away these other planets' irregular orbits, as Ptolemy did, Copernicus imagined an entirely different scheme—a solar system in which the planets revolved around the sun. Like Aristotle, Copernicus could find truth only by reinventing his conception of the universe and challenging the assumptions of his time.

Einstein accomplished a similar feat not just once, but time and again. His work continually redefined our picture of the universe, extending our understanding of phenomena on both the cosmic and atomic levels. To formulate his special theory of relativity, for instance, Einstein imagined an entirely new kind of universe—one in which gravity played no role. Although this view of the universe isn't accurate, it allowed Einstein to identify for the first time the rules governing light. To do so, he had to ignore Newton's laws of gravity, which say that every body exerts some gravitational pull on every other body. Einstein's method doesn't seem like a recipe for success. After all, why should a model based on an inaccuracy predict accurate rules governing the universe? But just like Aristotle and Copernicus before him, Einstein challenged the accepted beliefs of his time and, as a result, dramatically sharpened our understanding of the universe.

Just like Aristotle and Copernicus before him, Einstein challenged the accepted beliefs of his time.

Hawking's first famous achievement grew out of Einstein's theory, and like Einstein, Hawking angered a lot of people with it. Perhaps no one was angered more than Einstein loyalists. Hawking's assertion that the universe formed out of a big bang singularity created a problem for general relativists (Einstein devotees), since the laws of general relativity don't hold at singularities. The notion of a singularity is relatively easy to picture, but its properties are difficult to understand fully. Singularities are infinitesimal points, pinpricks in space, that are so dense that spacetime curves around them. Anything close to a singularity will be pulled toward it at a violent rate. Hawking's theory is dramatic because it reverses the existing theory of singularities. If singularities can attract matter and energy to themselves, he asked, why can't they do the opposite?

Why can't singularities repel matter and energy at ferocious speeds? In real time, they can't. But Hawking did something with the model of singularity that no one had ever imagined doing before: he looked at it in backward time.

This notion of backward, or reversed, time figures prominently in Hawking's book. It sounds much more abstract than it actually is. Simply imagine any object that is undergoing some continuous motion—for instance, a woman walking from her office to her home. In "normal" time (or, as we're accustomed to calling it, simply "time"), we observe the woman crossing streets, waiting at stoplights, and pausing at a newsstand to buy, say, a blueberry lollipop. We could easily document her actions by drawing a graph with one axis representing her position in time, and the other representing her position in space.

If we reverse the time, however, we have only the image of the woman standing on her front porch and being greeted by her husband—but neither we nor her husband know what events brought the woman there. Assuming that the woman started at her office gives us a starting point but doesn't help us know exactly what path she took home. So we must work backward. After greeting his wife with a kiss (and an intimate one at that), the husband notices a distinct taste of artificial blueberries. So his wife must have stopped at a candy store. Imagine that there are only three such stores in the town—so now we know that her route must have passed through one of them. Realizing that only one of these stores sells blueberry lollipops, the husband can make an even more accurate guess about his wife's route.

Astrophysically speaking, Hawking is a very attentive husband. He interviews every other pedestrian in the town, he times each stoplight, and he inquires about whether the town has jaywalking laws (and whether they're enforced). Most important, he knows the wife's route to work, which she follows religiously. There's no reason, at least not any mathematical one, why she shouldn't follow the same path on the way back home.

We might therefore think of reversed time as an investigation into the past, much like the process of finding a criminal based on the clues he left at the scene of the crime. Hawking's first significant scientific breakthrough came when he followed time backward from its current state and examined the possible routes it could have taken to get to what it looks like today. He concluded that if the fundamental tenets of general relativity hold and if the universe is indeed expanding, then the universe must have begun with a big bang singularity.

Once we accept this statement as true, we're lead to another breakthrough: now we can study the big bang without having to rely on what little evidence we have of it. Instead, by studying what happens at black hole singularities, we can learn about what happened at the first singularity—the big bang. This idea is quite convenient, since the big bang is very far away in both space and time, whereas black holes are all around us, in our own cosmic backyard. With this new information, we can get much closer to determining what existed before the big bang, whether the universe will keep expanding forever or if will collapse upon itself, and why it smells like artificial blueberries.

CHAPTERS 4–5
QUANTUM MECHANICS AND UNCERTAINTY

The most interesting thing about singularities is that they're the one major exception to the biggest rule of astrophysics—general relativity. Black holes, and especially the big bang, are often imagined as gigantic, cataclysmic events. But their size often gets confused with the power of the forces they exhibit. It's important to remember how small a singularity actually is. It's so small that when observing it or theorizing about it, we have to take into account the other great theory of physics—quantum mechanics.

Quantum mechanics is the name for the set of theories that describe the smallest phenomena in the universe. These theories describe events that happen on the order of the molecule or atom. To understand the theories of quantum mechanics, it's crucial to know that they all grew out of a single, monumental theory called the uncertainty principle. This principle states that we can't know both the exact location of a particle and its exact velocity at the same time. As Hawking discusses at the beginning of Chapter 4, our methods for determining a particle's location directly affect the particle's velocity. The more accurately we can determine the one variable, the less accurately we can determine the other.

The uncertainty principle has many implications for physics. Perhaps the most famous of these is the development of the dual theory of the light photon, which says that photons act as both waves and particles. But even more important for Hawking, the uncertainty principle has important implications for all science.

The German physicist Werner Heisenberg first articulated the uncertainty principle in 1926. The mid-1920s were marked by accelerated scientific progress. Ernest Rutherford had just predicted the existence of the proton and the neutron, Alexsandr Friedmann had just released his models for an expanding or oscillating universe, and Einstein's theories of relativity had gained widespread acceptance. It seemed that scientists were quickly approaching a complete understanding of the physical world, close to reaching the ultimate goal of science—self-negation. For if our understanding of the world was complete, there would be no more questions for science to answer. But into this optimistic and fruitful era the uncertainty principle was introduced. It was as if nature were striking back at human intellect. Science seemed as if it had reached a dead end. We could attain an understanding of the world at smaller and smaller levels, past the atom and even the photon, but we could never go further. Science had proved its own limitations.

"The question remains . . . how and why were the laws and the initial state of the universe chosen?"

The uncertainty principle destroyed humankind's fantasy of being omniscient one day. It forced many scientists to agree with priests and skeptics alike that science, indeed, could never fully explain our world. Interestingly, although Hawking acknowledges this limitation of science in this section of the book, he glosses over it later in the book's hopeful conclusion. But the fact is that as close as we might get to determining the true nature of our universe, Heisenberg's principle says that we'll always encounter some degree of uncertainty in our research.

This isn't to say that Heisenberg's innovation has no scientific value. In fact, it accomplished the opposite: it paved the way for many of the most significant scientific advancements made in the decades since, including many of Hawking's own achievements. Nevertheless, as much as the uncertainty principle has given us a greater understanding of the universe, it also predicts that we can never achieve a complete understanding. It also raises an unsettling question, one that nags Hawking throughout his book: if there's one dead end in science, why can't we assume that there are others?

Indeed, in 1949, a second dead-end principle was formulated—mathematician Kurt Gödel's "incompleteness theorem." This theorem predicts a breakdown not in physics but in the fields of logic and computer science. Although Hawking claims that neither Heisenberg's nor Gödel's theories have forced us to lose hope for a unified theory of the universe, the theories occasionally rear their ugly heads, tempering Hawking's optimism a bit. What gives Hawking and other astrophysicists hope is that some scientific theorems that were once considered dead ends have in recent years been revisited and, with the help of new theories, resolved.

The most famous example is Einstein's theory of general relativity. Scientists have realized that Einstein's theory predicts its own failure. It was very successful in predicting what would happen in situations involving bodies that have large mass, like stars and planets. However, it faltered in its prediction that there should be points of infinite density in the clumpier parts of the universe. Although these points do exist—we call them black holes—they aren't governed by the force of gravity, as Einstein's theory predicts. Instead, because these points are infinitely small, they're governed by the forces that affect small particles. Since quantum mechanics describes these small-particle forces, we must use quantum mechanics to describe black holes. Einstein didn't incorporate the rules of quantum mechanics into his theory, so his theory is incomplete. But Hawking is one of several theoretical physicists who have begun to apply the rules of quantum mechanics to the study of black holes, and he has therefore helped to combine quantum mechanics with

If there's one dead end in science, why can't we assume that there are others?

Einstein's theory. Consequently, Hawking has made significant breakthroughs in our understanding of black holes. One such example is his discovery that black holes are not, in fact, black.

CHAPTERS 6–7
BLACK HOLES

Despite its age (it was only coined in 1969), the term "black hole" may be the most familiar term in all of cosmology, maybe even inching out "big bang" for this distinction. But this isn't to say that black holes are well understood. Like many concepts in cosmology, a full understanding of black holes evades most laypeople. Our current picture of black holes didn't surface until the landmark work of Hawking and his colleague Roger Penrose in the mid-1970s.

When we think of black holes, our first impulse is to imagine a small region in space that sucks in everything around it. Even light is unable to escape the enormous gravitational pull. We picture a singularity, a small point of infinite density, at the black hole's center. This density is what provides the strong gravitational pull. Newton's theory of gravity, which says that the attraction between objects depends on their masses and the distance between them, helps explain such a pull. The larger the masses and the shorter their distance from each other, the more attracted two objects will be. A singularity has such an enormous mass that any object within a certain distance, no matter its size, will be pulled toward it.

With Newton's theory in hand, we understand why particles near this singularity would be pulled toward it and why particles far away from the singularity would not feel its gravitational pull (after all, we ourselves haven't been sucked into some faraway black hole). But it's difficult to imagine what happens between these two extremes. Imagine a particle near the imaginary line, called the event horizon, that separates the black hole and the rest of the universe. The particle is a little like a mouse ambling right in front of the jaws of a sleeping cat. How can the gravitational force be so powerful on one side of the event horizon yet disappear entirely on the other side of the horizon? The basic answer is that just past the event horizon, the pull isn't immediately infinitely strong but only *begins* to be felt at the horizon. The gravitational pull slowly takes up the particle before the particle rapidly and exponentially increases speed until it is crashes right into the singularity.

But the presence of an event horizon still produces a very odd environment in the immediate vicinity. Since particles move randomly, many of the particles close by an event horizon will likely move randomly within it. The region closest to the event horizon will therefore be greatly depleted of particles. Other particles may randomly drift into the event horizon's neighborhood, but many of those end up either moving back out of it or moving further, past the event horizon. So the region close to the event horizon is nearly empty. In fact, it's so close to being empty that the only reason we can be sure that any particles are present is because of the uncertainty principle. If there were no particles, this empty region wouldn't have any electrical charge and wouldn't change over time. The uncertainty principle says that we can never be certain of both a region's charge and rate of change—but if both charge and

"The number of black holes may well be greater even than the number of visible stars, which totals about a hundred thousand million in our galaxy alone."

change were zero, we'd know both. Therefore, the uncertainty principle forces us to infer minimal charge and minimal change, which can be imagined as brief fluctuations in space, like blips on a radar screen that appear and disappear instantaneously. These blips are an exceptionally odd phenomenon.

A second space oddity is the fact that black holes are seemingly incompatible with Einstein's theory of the conservation of mass. This is the theory expressed in the famous formula $E = mc^2$, which states that the energy of a system is equal to its mass multiplied by the speed of light squared. If the mass in a system decreases, then the energy of that system must increase somehow. Mass can't disappear without something else taking its place.

Yet that's exactly what the old theory of black holes seemed to predict. Of course, we might try to formulate a provisional excuse for this phenomenon. Since black holes curve space-time to such a great extent, we theorize that the laws of physics don't hold inside them. The conditions inside a black hole don't resemble those in the rest of the universe. For example, there's no such thing as time inside a black hole. What happens inside a black hole is entirely unpredictable. So why should the law of conservation of mass be in effect?

Nevertheless, this logic still doesn't hold. If it did, then all the black holes in the universe would be sucking up particles and light, and the known universe would gradually lose mass and energy. The laws of mass and energy conservation wouldn't be accurate. This old model of black holes ultimately predicts a depletion of energy and mass in the universe.

Hawking's solution to this problem provides a good example of his way of thinking, exemplifying his ability to question the sturdiest beliefs of his time. In the black hole problem, Hawking was confronted with two prevailing—and mutually incompatible—models of the universe. The prevailing definition of black holes contradicted the conclusions of the theories of mass and energy conservation. One of these two theories, or both, had to be at least partly inaccurate. Hawking's solution to the problem helped us to define the black hole more accurately and led to a number of other discoveries that changed our conception of the universe.

Hawking's solution took into account both the oddities of the "empty space" surrounding a black hole and the problem of black holes' draining the universe's energy and mass. Hawking realized that the small fluctuations of energy and mass in empty space at the event horizon were in fact particle and antiparticle pairs appearing and just as quickly annihilating each other. This concept is a difficult one to imagine—how do particles just appear, we might ask. But the phenomenon is predicted by the uncertainty principle and by some of the more obscure tenets of quantum theory (along with another factor that not even Hawking fully attempts to explain). What's important to understand is that when these particles appear exactly at the border of the black hole—at the event horizon—sometimes they're unable to pair and annihilate each other. A particle with positive energy might fly away from the black hole while its partner falls in. In this case, we can observe what looks like light, or radiation, being emitted from the black hole.

And thus Hawking's grand conclusion: "black holes ain't so dark." With this discovery, Hawking reshaped our perception of black holes. If that term weren't already so popular, we might be tempted to rename them "white holes." Even more impressive is how Hawking used his discovery as a stepping stone to a far more profound theory—one that involves more than just black holes and forces us to question our conception of the universe.

CHAPTER 8
THE BEGINNING AND END OF THE UNIVERSE

Hawking is at his best when he elucidates the most difficult concepts of science. Unfortunately, he's not always consistent in his technique. In some instances, he walks us through a theory, explaining at each step how it was conceived and then showing us how this theory led to another. But sometimes—usually with shakier theories that require mathematical proofs instead of simple explanation—he tends to take us through shortcuts. Take, for instance, the statement (and take a deep breath now), "[O]ne can calculate the probability that the universe is expanding at nearly the same rate in all different directions at a time when the density of the universe has its present value." Although we might expect a reasoned explanation, Hawking never delivers. The only explanation would appear in the math, and Hawking isn't about to teach us higher calculus, so we're forced to trust him. Perhaps the most frustrating instance of this shortcut technique is Hawking's discussion of "imaginary time," a solely mathematical concept. Don't try to think of it as anything else—your mind might start reeling. The concept of imaginary time is useful because many complicated mathematical operations about the universe turn out answers that are expressed in imaginary numbers.

Hawking wasn't dismayed when his calculations included imaginary numbers. Instead, he used them to shape a new, valid portrait of the universe—even though this new picture of the universe is somewhat threatening to the idea of a divine creator. With the help of imaginary numbers, Hawking proposed a universe with no boundaries, no beginning or end. In such a conception, singularities are not viewed as endpoints in time and space, and even time itself has no direction. Hawking proposes to use a sphere as a metaphor for thinking about the universe. The universe becomes a complete system, having no more beginning or end than Earth does.

Hawking's findings about black holes provide more tangible examples of how this no-boundary theory works. We've discovered that black holes don't, in fact, act completely like maverick chunks of space that refuse to behave according to universal laws. Black holes do remove energy from the universe, but they also create energy out of nothing—from the empty space surrounding them. So we can say that black holes *do* obey the law of energy conservation. Whatever happens within the borders of event horizons, black holes aren't vacuums—they create just as they consume. They mark neither a beginning nor an end. And since we believe that our

"[I]f the universe is really completely self-contained . . . it would have neither beginning nor end: it would simply be. What place, then, for a creator?"

universe began with a singularity, we now have to question how to characterize this beginning, or even whether or not we can accurately describe it as a beginning. If singularities aren't endpoints in time, then the big bang can't mark a real beginning of the universe.

CHAPTERS 9–10
TIME'S ARROW AND WORMHOLES

Time travel is such a fascinating subject that Hawking added a whole new chapter on it for the updated edition of his book. Although there have been new theoretical advances in the investigation of time travel in the past twenty years or so, the possibility of time travel remains relatively hopeless. Hawking's theory of a no-boundary universe, however, can help us imagine what time travel might look like.

Again, the model of a sphere—or more specifically, of Earth—is helpful. We can imagine starting at the North Pole as time begins and walking down the side of the sphere, though we perceive our path as only a slope, not a straight line. At our present state of the universe, we've hiked as far down as, say, Boston. The circles of latitude have become wider as we've traveled south, just as the universe has expanded as time has passed. But we don't have the ability to turn around or walk backward. Nor do we have any sense of space above us or below (underground)—above and below represent dimensions of space-time that humans can't perceive.

Since the universe is still relatively young, imagine our imaginary hiker as a young child. In Boston, he walks into a park that has a sandbox, picks up a shovel, and begins to dig. He's moving in an unprecedented direction, and it takes an immense amount of energy. But after days and days of digging, the child finds that he's begun to go down deep into the Earth. After many more days, he's surprised to see a light at the end of his sand tunnel. He continues digging and finally ends up in another sandbox—in Beijing. He's fulfilled the old schoolboy myth of digging right through the Earth until he reached China.

Such a feat is possible because the Earth, like space-time, is round—if you spend energy to go into it, you'll end up elsewhere. What's more, the trip, as long as it took, was a more direct path to China than if the boy had walked there by foot. It was a shortcut. Similarly, if the child's tunnel were a wormhole, he would have gone to another time more quickly than it would have taken time itself to get there. Going through the tunnel from Boston to Beijing, he would have traveled to the future; a Chinese boy digging in the opposite direction would have traveled to the past. The problem with wormholes, though, is that it they require an incredible amount of energy, far more than all of humanity can muster.

CHAPTERS 11–12
CONCLUSION

Hawking's conclusion might be surprising coming from a theoretical physicist, but it explains much of the popularity of *A Brief History of Time*. As we know long before the book's end, the grand goal of science is to find a theory of everything, a theory that predicts every possible phenomenon of the universe, including its beginning and end. Hawking writes that there are currently many partial theories of the universe, each mastered by a small number of specialists. But no one has yet managed to combine them in any complete way. Considering the highly advanced concepts involved in this area of cosmology, we might expect a theory of everything to be utterly arcane, restricted to scientific minds the likes of Hawking and few others. Or perhaps such a theory would be so complex that it would be incomprehensible to any human mind.

But Hawking's prediction is surprisingly populist: he predicts the exact opposite. "If a complete unified theory was discovered, it would only be a matter of time before it was digested and simplified in the same way and taught in schools," he writes. "We would then all be able to have some understanding of the laws that govern the universe and are responsible for our existence." The conclusion of the book reminds us why Hawking is so popular. He believes in the popularization of physics, of knowledge. He is not only an excellent physicist but also an excellent teacher, for his ultimate goal is not just advancing science but helping humanity to share these advancements.

Hawking never comes across as condescending. At the heart of his drive toward scientific knowledge is a genuinely felt compassion for

Our own worst enemy

The biggest challenge facing humanity today, Hawking believes, is our "aggressive instincts." In an interview with Larry King, he explained, "In caveman—or caveperson—days, these [instincts] gave definite survival advantages and were imprinted in our genetic code by Darwinian natural selection. But with nuclear weapons, they threaten our destruction." Hawking also believes that we face extinction in the next one hundred years unless we begin to colonize space.

humanity, one that compels him to share the word of science with as many people as possible. This motivation justifies Hawking's approach, his attempt to simplify the most difficult theoretical concepts to an easily understandable level.

What first manifested itself as an inquiry into the existence of God becomes an inquiry into the nature of man. Hawking realizes that our scientific quest to determine the possibility of a divine creator of the universe tells us more about ourselves than about a supreme being. If man could someday articulate a true theory of everything, it would not be proof of a deity but rather the highest assertion of mankind's intelligence. It would be humanity's greatest possible achievement. So although Hawking begins with an inquiry into the existence of God, he concludes with the realization that our goal is actually "a complete understanding of the events around us, and of our own existence."

POINTS OF VIEW

Take the Cosmos Quiz

Mathematical models have explained many bizarre phenomena, but Hawking's looking for the one rule to ring them all.

○ ○ ○

Why is Newton such an important figure even today? What did he contribute to cosmology?

LEARN NEWTON'S LAWS, PREDICT THE FUTURE!

When we think of the most famous physical scientists in history, Newton may be only one of several who come to mind. Aristotle, Pythagoras, Ptolemy, Copernicus, Brahe, Galileo, Kepler, Schrödinger, Einstein, Heisenberg, Planck, Bohr, Dirac, Feynman, and Hawking would certainly all be members of this elite club. But if we had to pick an even shorter list—the elite of the elite, as it were—we would probably be reduced to about five of these figures. If we were really pressed, the list would probably shrink to about two—Einstein and Newton. (And, perhaps, God.)

When Hawking describes Newton's *Principia Mathematica* as the most important work in physical science, he's not merely heaping praise on an idol—he's stating a fact. Newton's laws of motion, which he advances in *Principia Mathematica*, describe and predict the actions of all forces and bodies. This is no small feat. Newton's work dwarfs the greatest discoveries made by virtually any physicist since. Many modern physicists have attained greatness by making findings in extremely specific, often arcane fields of study. But Newton was the first to comprehend a huge chunk of the cosmic plan. What's more, he also developed the complicated mathematical frameworks that prove his theories.

Newton's findings center on three laws, each of which can be summarized in a single statement. Newton's first law, the law of inertia, states that a body remains at rest, or moves in a straight line at a constant speed, unless acted upon by a net outside force. The implications of this law are crucial. All movement is caused by the application of certain forces. Therefore, we can express the movement of a body by the forces that act upon it. Objects don't have energy within themselves. If a ball stops rolling, then some force made it stop—it can't have simply "run out of steam" or lost some sort of internal energy. The ball stops because the force of friction (the floor rubbing against the ball) overtakes the force of the original push a person might have given the ball. The only way to keep the ball moving at the same speed is to continue to push the ball at that speed.

Newton's laws allow us to understand and be able to predict virtually every action in our known universe.

The words "net outside" are crucial here too. Newton recognized that any body in action is subject to many forces that dictate its movement. When Newton applied his first law to celestial bodies, he determined that some force had to be acting on the planets continuously—if not, the planets would fly off into the far reaches of the universe rather than stay in orbit around the sun. This force, he realized, was gravity.

Newton's second law states that the net outside force on an object equals the mass of the object times its acceleration. The greater the force, the more the object accelerates. The larger the mass of an object, the more force is needed to make the object move faster (or move at all, if the object begins at rest). This law is easy for us to picture: it's easier to throw a baseball than a bowling ball. Also, the harder a ball is thrown, the faster it goes. As intuitive as this may seem to us today, the mathematical implications of the theory are profound. For one, it helped Newton to determine the exact force of gravity on Earth—9.8 meters per second squared, regardless of the mass of the object. This figure, in turn, is essential for predicting the motions of all universal bodies (except those of small particles, which are governed by the laws of quantum mechanics).

Newton's third law: "Whenever one body exerts a force on a second body, the second body exerts an equal and opposite force on the first body." More succinctly, for every action there's an equal and opposite reaction. When we push against a wall, the wall exerts an equal force back against our palms.

Although the third law may be slightly difficult to rationalize, it led directly to a fourth law, which is known as Newton's universal law of gravitation: "Two bodies attract each other with a force that is directly proportional to the mass of each body and inversely proportional to the square of the distance between them." This is the big law. It tells us that *everything* in the universe exerts a force on everything else. The sun exerts an attractive gravitational force on the Earth, which keeps us in orbit. But the Earth also exerts an attractive force (though a significantly smaller one) on the sun. You exert a force on the person sitting next to you, and they exert one back on you.

Newton's laws of motion and his universal law of gravitation allow us to understand and predict virtually every action in our known universe, from the movement of a ball being hit by a bat to the movement of galaxies. Newton's laws are the rules to which space oddities like singularities and antiparticles are the exceptions. In this sense, Newton's contribution remains essential to our understanding of the universe. Newton's laws are also the template Einstein used to develop his theories of relativity, which might be imagined as extensions of Newton's laws.

Any great theory of everything will be indebted to Newton's theory—or will evolve out of his theory. His laws describe what we might call all "practical" events, those that we encounter and observe in everyday life, both on Earth and beyond. Today, we still rely on Newton's laws in every field of engineering or design, whether it be for making spaceships faster or cars safer, buildings more structurally sound or sports equipment more effective. But this kind of summary of the uses of Newton's laws is only an understatement. They describe our every action, both those we can see and those that are too small—or too vast—for us to perceive.

What does Einstein's theory of relativity mean for us? What's space-time? How can it be "curved"?

TAKE OUT A PIECE OF PAPER . . .

Einstein's theory of relativity is undoubtedly one of the most important theories in physics since Newton's laws of motion and gravitation. In fact, Einstein's ideas are an extension of Newton's theory. Newton imagined gravity as simply a kind of force. This is not altogether incorrect—after all, physicists still talk about a "gravitational force," which is used to calculate events that happen in our everyday world. But this idea didn't satisfy Einstein. He realized that extremely massive objects, like the sun, actually manage to change the way space and time act around them.

To understand Einstein's idea, we must define the term "space-time." Put simply, space-time is a combination of the three space dimensions (length, width, and breadth) and the time dimension—a way of pinpointing an object's location in both space and time at once. One way of imagining space-time is to picture a single slice of it at a time, as if it were a sheet of paper, for instance. This sheet represents a single moment in time (we are imagining two, rather than three, spatial dimensions for greater simplicity). A speck on the sheet of paper represents a certain point in space at a certain point in time. A second later, we stack another sheet of paper over the first one. On the surface of this second sheet is a speck in a slightly different location. It has moved in both space and time. Space-time is what we get when we create an infinitely large stack of paper.

In a sense, space-time is a way to think about both space and time at once. Einstein discovered that space and time actually do behave like a flat plane, like a sheet of paper. Normally, as on the Earth, space-time is flat. When we walk in a straight line, we move in a straight line. Similarly, time elapses in an orderly, linear fashion. These ideas sound obvious, but around large gravitational pulls, time and space cease to be flat. The planes bend or curve. An enormous object, such as the sun for instance, can be imagined as a rock on the sheet of paper. It causes the sheet to sag from the weight. In the same way, the dimensions of space and time become warped. If we decide to walk in a straight line past such objects that bend space-time, our path won't resemble what we normally think of as a straight line—we'll dip down and walk on an angle. It's much like walking in a straight line on the slope of a valley: we're walking straight (feet perpendicular to the ground), but to someone standing on the flat bottom of the valley, it will appear as if we're walking inclined at an angle. When the dimensions of time and space are traversed at an angle, some strange things happen.

The first major application of general relativity was to predict the orbits of the planets around the sun. Newton's law came close to describing the motion of planets, especially those farthest from the sun, but it did a poor job of predicting the orbits of the nearer planets. In the area right around the sun, space-time is extremely warped, or curved, so paths that pass near the sun become curved as a result. In the cases of planets far from the sun, general relativity equations predict virtually the same orbits that Newtonian equations do, since space-time is almost completely flat in these regions. The less gravity present, the less curvature there is to space and time.

The general theory of relativity also tells us something about the speed of light. Einstein's theory helps us understand why the speed of light is constant. What's more, Einstein realized that light moved at the same speed (3×10^8 meters per second) regardless of where the observer stood or how fast he was moving in respect to the light source. This idea came from observation rather than theory. But Einstein was the first to trace mathematically and theoretically all of the implications of this fact, including bizarre phenomena such as watches slowing down and rulers shortening at speeds close to the speed of light.

○ ○ ○

Why do we think that the universe began with a big bang? What evidence do we have that a big bang occurred?

THE AMAZING EXPANDING UNIVERSE

The theory of the big bang, as we now know it, began with Edwin Hubble's observation that all stars seemed to be moving away from us. Moreover, he noted that the farther away the stars were, the faster they retreated. At first, some physicists believed this observation implied that we are at the center of an expanding universe—a startling return to the geocentric views of ancient times. But the scientific community soon realized that stars are not only moving away from us, but also moving away from one another. If we trace this expanding universe back in time, we ultimately reach a point in the past when everything was condensed into a single point. The density of this point would have had to be extremely high, and it could have caused an explosion of sorts—the explosion we call the big bang.

Hawking's great innovation was to argue that the point of the big bang was in fact a singularity, the same kind of dense point that is at the center of all black holes. Although it seems intuitive that a highly dense point could explode as a big bang, Hawking was the first to prove that this absolutely should be the case. Though he later amended this point in his model of his "no boundary" universe, his logic is clear and still holds. Roger Penrose had already proven that "any body undergoing gravitational collapse must eventually form a singularity." Hawking simply reversed this theory, reasoning that singularities, then, must inevitably explode. He then viewed the entire universe as the product of an explosion and proved mathematically that an expanding universe must have begun with a singularity. Even though the idea makes sense in theory, only highly rigorous mathematics supplies Hawking's proof. So on this matter, we're forced to trust him.

The idea of an expanding universe is difficult to grasp, especially since the universe doesn't expand in the traditional sense of the word. The typical question regarding this model of the universe is usually something like,

"If the universe is expanding, what is it expanding *into*?" But this question implies that the universe has some end, which it doesn't, in fact, have.

The balloon metaphor that Hawking employs is the best way to imagine this expansion. The universe is like a balloon—when it expands, it inflates. Inflation is actually a more appropriate word than expansion, and indeed, inflation is used to describe the first stages of the universe (after the big bang, we say that the universe went through a very brief, but very powerful, inflationary period). We shouldn't imagine the expanding universe as an army pushing past some boundary, claiming new land. It's more like a balloon that inflates, making the points on its surface grow farther and farther apart. The universe is elastic, so the space, the distance between things, grows bigger and bigger. This is true only in large regions of space. It isn't true within bodies or even within structures as large as galaxies. The gravitational pull between things like stars and planets, or planets and people, prevents the universe's expansion energy from stretching apart the actual bodies themselves. The Earth isn't slowly getting larger, nor is the universe. In other words, don't try to blame your "expanding" waistline on the big bang.

○ ○ ○

What's the uncertainty principle and why is it significant? Why does it affect our understanding of the universe if it applies only to the study of the smallest particles?

WE THOUGHT WE WERE SMART, BUT THEN . . .

The uncertainty principle states that we can't accurately measure both a particle's position and its momentum (velocity) at the same time. The more accurately we measure one of these variables, the less accurate the other measurement will be.

But why is it necessary to be able to determine both of these variables simultaneously in the first place? The answer lies in our very definition of a scientific theory. A scientific theory is useful only if it predicts some future situation based on present circumstances. Newton's theory of gravity is proven, in part, by the fact that the theory successfully predicts that every time an apple becomes disconnected from its branch, the apple will fall until it hits something below (the ground, for instance, or someone's head). Any theories about elementary particles must make predictions about the particles' future states based on present conditions.

For example, take the problem of predicting the distance to a star. We receive light particles from the star, which travel to us at a certain velocity and at a certain angle. To determine the star's location, we must study both the speed and location of these light particles so that we can then trace them back to the star. But to do so precisely, we must know the particles' exact velocities, which is the type of measurement the uncertainty principle declares to be impossible. We can figure out an average velocity of the light particles at some general location in order to make a fairly reliable prediction about the star's location. However, we can only guess the star's location within a certain area, give or take a certain number of light-years. This prediction is neither accurate nor verifiable.

In a field like quantum mechanics, in which virtually every experiment depends on us knowing a particle's location and velocity, it's easy to see how the uncertainty principle could wreak a lot of havoc. Quantum mechanics is an imperfect field for exactly this reason. Its theories don't predict specific outcomes but a group of all possible outcomes, which are then ordered according to likelihood. But, you say, how does this affect astrophysicists? After all, there are other ways of determining the location of a star (by analyzing the star's gravitational effects on stars around it, for instance, or by comparing its brightness with those stars whose location we already know). Who cares what the particles are doing when we know what the star is doing?

Hawking, for one, cares deeply. He strongly believes that the long-coveted theory of everything will combine both the quantum theory and the theory of relativity. As a result, we'll have to apply the uncertainty principle to cosmic events that were once thought to be strictly within the domain of relativity theory. Hawking's greatest achievements have been in showing how these two grand theories can be combined. His theory that black holes emit radiation was the first time that the uncertainty principle was applied to a large-scale cosmic event—the black hole. Hawking realized that to analyze the event horizon—the microscopic border between a black hole and the universe around it—we had to employ quantum theory. The theory of relativity, which describes the vast machinery of space, isn't applicable on such a small scale. Hawking needed to zoom in on an infinitesimal place in space, and so he had to use the uncertainty principle to understand how particles acted there. As a result, he made a revolutionary discovery.

The singularity that preceded the big bang is also an infinitesimal point. To understand the conditions that led to the big bang, we must apply the uncertainty principle to this singularity. As is often the case with the uncertainty principle, we must use the unpredictable to make a correct prediction.

○ ○ ○

Do antiparticles really exist, or are they just a cool way of describing an abstract theory? How can we imagine "antiworlds" and "antipeoples"?

IMAGINE ALL YOU WANT, BUT DON'T TOUCH

Theoretical astrophysicists often feel the need to make abstract, mathematical concepts sound like some 1950s science-fiction flick. They are science's greatest pitchmen. "Imaginary time," "space-time," "virtual particle," and "antiparticle," are all physics-speak for theoretical and abstract things, concepts that without these names might bring us to tears of boredom rather than to tears of glee or terror.

Antiparticles do exist, and have been proven mathematically if not by direct observation. But to understand them, we have to make a leap of logic that's not intuitive. The uncertainty principle guarantees that, over a brief enough interval of time, particles can spontaneously appear and disappear. This is something we have to accept. The more matter that appears, the less time will elapse before it disappears. This implication of the uncertainty theory is not easy to explain without resorting to Heisenberg's calculations. This spontaneous appearance and disappearance of matter is guaranteed by the mathematical expression of the uncertainty principle, not by any conceptual reasoning. Unless we want to dig through the math, we have to trust Hawking (via Paul Dirac) on this one.

But particles can't just appear by themselves. To preserve the balance of energy in the universe, an antiparticle appears for every particle that appears. The antiparticle is identical to the particle in virtually all respects except for its electric charge. To simplify the situation, imagine the charge of a particle is +1 while that of an antiparticle is –1. Since (+1) + (–1) = 0, the total energy of the universe stays the same—the positive energy cancels out the negative energy. The universal law of conservation of energy still holds. Although particles come and go, the energy of the universe is preserved.

For a more specific case, imagine an electron appearing spontaneously in space. An electron, one of the particles that orbit the nucleus of an atom, has a charge of –1. Therefore, when an electron springs into existence, a particle with a +1 charge will come into existence too. This positively charged "electron" is an antiparticle, since its charge is the opposite of the regular particle's charge. This positively charged particle is called a positron. The antiparticle that would accompany the spontaneous appearance of an entire atom would consist of a nucleus of antiprotons (carrying negative charges) orbited by positrons (carrying positive charges) that in all other respects would act just like a normal atom.

If you want to get fancy, as Hawking does, you could extend this to an increasingly larger scale by imagining antimolecules and anticompounds, antipeoples and antiworlds. All matter can be reduced to tiny particles of either positive or negative charge. The positives balance out the negatives in a way that allows matter to keep existing. If the positives and the negatives were reversed, nothing would appear to be different. It would be like watching a basketball game between two equally matched teams, in which the players suddenly swapped jerseys with their counterparts on the other team. The game wouldn't be played any differently—we would just call the players by different names.

For any antiuniverse to exist, however, it would have to be quite far from our normal universe. As soon as antiparticles come into contact with regular particles, both spontaneously cease to exist. So, no, you shouldn't shake hands with your antiself. But the truth is, he or she would never even make it to your front door.

What are virtual particles? Do they really exist, or are they just a theoretical concept people use to sound really smart?

PHOTONS, GLUONS, AND BOSONS . . . OH MY!

In short, yes, virtual particles do exist, but, yes, people often talk about them just when they want to sound smart (or when they're hallucinating). A virtual particle is particle that we can't detect directly but that is emitted from an observable, or "matter," particle. Virtual particles also carry forces that act on other particles in observable ways. This last definition requires a bit more background, and a bit more history—especially since even high school science classes teach physics as if there were no such thing as virtual particles.

"There could be whole antiworlds and antipeople made out of antiparticles. However, if you meet your antiself, don't shake hands! You would both vanish in a great flash of light."

When we're taught that gravity is a force, we usually inquire no further. We learn that we can quantify its pull, we can predict what makes it stronger or weaker, and we can describe how other objects act in its presence. But what about this force, what *is* it actually? A force is a difficult thing to imagine. We might take it, like imaginary time, for hocus-pocus, if we didn't feel it so strongly all the time. Still, the idea of one object exerting a pull on another object is difficult to grasp.

It was only in the 1940s that physicists figured out what forces are. Force, they found, is actually made of particles. But since these particles are so small and exist for such a short interval of time, even our best particle-detecting machines can't observe them. So scientists describe them with the deceptive but flashy term "virtual particles." The word virtual implies that the particles don't actually exist, when in fact they do—scientists just haven't discovered a machine precise enough to detect them.

Antiparticles, confusingly, are sometimes called virtual particles. But for the most part, the term virtual particles describes particles that carry force. These particles are gravitons (which carry gravitational force), gluons (the strong force), photons (electromagnetic force), and intermediate vector bosons (the weak force). Combined, these forces control the interactions of virtually everything in the universe.

Some of the names for these particles sound absurd, but their functions are relatively easy to describe. Take the graviton, for instance. Every atom of every object releases gravitons, firing them off into space. The act of firing off gravitons causes the atom to recoil, the same way a person might recoil after firing a shotgun. The bullet, or graviton particle, hits another atom and is absorbed, changing the velocity of that second atom. Many gravitons are exchanged in a very short time, and soon enough, the two atoms are headed toward each other. From our vantage point, we notice a gravitational pull.

The other virtual particles work in similar ways. In fact, most scientists believe that all four forces are components of one greater force. So far, physicists have figured out that at extremely high energies, three of the forces (all but gravity) resemble each other exactly. But at normal energy levels, the forces are rather different. The strong force works on an extremely small scale—at the level of the atom, holding protons and neutrons together. The weak force works at a slightly smaller scale, producing radioactive decay within atoms. The electromagnetic force is very strong, supplying the force between particles with electric charge. Virtual particles are these forces' tiny but energetic foot soldiers.

What will happen to the universe in the future? Will it destroy itself or exist forever?

STAY IN YOUR SEATS, WE'VE GOT A WAYS TO GO

The universe is expanding, and right now there's a near-consensus in the scientific community that it will continue to expand forever. There are two explanations for this expansion, though Hawking all but ignores one of them.

The basic calculation that predicts the future of the universe is based on the quantity of matter that exists and the universe's current speed of expansion. The latter is fairly easy to determine. We can measure how fast stars are moving away from us or from one another, so scientists have determined the universe's rate of expansion relatively accurately. The amount of matter in the universe is more difficult to calculate, but that can be approximated.

According to current models of the universe (first articulated by Alexander Friedmann), the universe will do one of three things, depending on the ratio of the universe's mass to its expansion energy: the universe will expand forever, cease expanding and collapse, or remain in a steady state. If there's more than a certain amount of mass in the universe, the gravitational attraction between all of this mass will cause it to drift together slowly. As a result, the expansion of the universe would slow down, and the universe would then begin to collapse. The universe would undergo the opposite of the big bang and expansion—a deflation leading up to a "big crunch." If there *isn't* enough matter to exert gravitational pull and overcome the energy of expansion, the universe will keep expanding forever. And if there is just enough mass to stop the expansion, but not enough to make it condense again, the universe will linger eternally at some given size.

The mass that scientists have observed in the universe isn't nearly enough to fulfill the requirements of the collapsing model of the universe, so the common consensus is that the universe will continue to expand forever. Hawking acknowledges this explanation and then proceeds to spend a great deal of time on his own theory of the no-boundary universe. Because this theory depends on the use of imaginary time, it has little bearing on our understanding of real time and the real universe.

Like imaginary numbers, this view of the universe manages to be provocative, despite being at its root no more than a mathematical innovation. This isn't to say that Hawking's idea of the no-boundary universe is useless. It helps us calculate certain things, such as the probability that the universe is expanding at the same rate in every direction. Nevertheless, Hawking's theory hasn't, at this point, significantly helped scientists to predict the future of the universe any better. Instead, by doing away with real time, Hawking's theory makes a universe with a future and a past irrelevant in the traditional sense.

Interestingly, this no-boundary model of the universe predicts that there's a high probability the universe will undergo a big crunch at some point. This idea contradicts the prevailing views of the universe—and even Hawking's former views that the universe will expand forever. How does Hawking justify this? He relies on the presence of "dark matter," which is matter that we know exists but that we can't observe directly. This unseen matter could be anything from huge deposits of interstellar dust to a mysterious new kind of matter that we haven't yet discovered. Scientists do have evidence that dark matter exists. Nevertheless, even if we added the mass of this estimated dark matter to the matter that's visible to us, we would come up with only ten percent of the matter necessary to halt the expansion of the universe. But since dark matter is so difficult to trace, it's possible that enough of it exists to force a universal collapse. In fact, this is the future that Hawking's model predicts.

Hawking has acknowledged the prevailing view of the future of the universe and has added his own idiosyncratic views to the mix. But he doesn't mention one of the most recent theories about the future of the universe. This new scenario primarily stems from observations the Hubble Space Telescope has made since the last printing of *A Brief History of Time*. Recent observations have shown that, shockingly, the universe's expansion is accelerating. These observations imply that there is a much greater source of dark matter than previously imagined, and this dark matter works in a peculiar way. Rather than add to the gravitational forces in the universe, this new kind of dark matter—or dark energy, as it's sometimes called—is repulsive. We still don't know whether this dark energy is caused by some kind of unknown form of matter that reacts against gravity or if it's simply a hidden fifth natural force, a kind of antigravity. It seems that this force, often called the lambda force, will create

Guess who's coming to dinner

Hawking predicts that if aliens ever visit Earth, they'll be more like the aliens from ***Independence Day*** than those from ***E.T.***: "As we explore the galaxy we may find primitive life, but not beings like us. Even if life develops in other stellar systems, the chances of catching it at a recognizably human stage are very small. It won't be a universe populated by many humanoid races with an advanced but essentially static science and technology. Instead, I think we will be on our own, but rapidly developing in biological and electronic complexity. It could be that there's an advanced race out there that's aware of our existence, but it's leaving us to stew in our own primitive juices. A more reasonable explanation is that there's a very low probability either of life developing on a planet, or of that life developing intelligence."

a universe that expands at increasing rates, as all bodies in space repel each other. The existence of the lambda force threatens Hawking's theory of a no-boundary universe. But we can leave the mystery of this new force to the Stephen Hawkings of future generations.

○ ○ ○

Is time travel possible? How could it be achieved? Could we make a time machine by fueling an '85 Delorean with uranium?

YEAH, BUT DON'T BET THE HOUSE ON IT

Time travel is possible—but just barely. At this point, the most probable method of time travel would be to go through a wormhole. A wormhole is a tunnel through space-time that connects to flat regions of space. The best way of imagining a wormhole might be to think of the world metaphorically: it's quicker for a worm to go directly through an apple (through a wormhole) than to crawl over it. Space-time, like apples, is curved. By traveling through a wormhole, we could move much faster than the speed of light. If we traveled even faster than light, the time measurement of our trip would be a negative number—so we'd actually be going back in time.

Unfortunately, wormholes are hard to find and even harder to create. For a wormhole to exist, there must be a large quantity of "exotic material," matter with negative energy density. This is a strange term, more easily understood by contrasting it to its opposite. Normal space has a positive curvature, through which isn't really possible to travel. For there to be a possibility of a wormhole, the universe would have to be bent the other way, in what Hawking calls a saddle shape.

Imagine that this book represents space-time. To get from the front cover to the back cover, we would have to climb over the spine of the book (or over the pages). Now take the book and hold only its two covers, letting the pages drop down (this is bad for the book, but good for understanding time travel). We've introduced negative energy density into our model of space-time. Now, the fastest route between the front cover and the back cover isn't a path over the spine, but a straight line between the two covers in space. This line would be a wormhole.

"The possibility of time travel remains open. But I'm not going to bet on it. My opponent might have the unfair advantage of knowing the future."

But it takes a lot of negative energy density to keep a wormhole open, and this, in turn, requires a lot of normal energy—much more than we currently have in our solar system, let alone on our planet. Without this energy, wormholes form and pinch themselves off almost instantaneously, forbidding the possibility of time travel. Nonetheless, it's impossible to predict what future civilizations will be able to accomplish, how much energy they will be able to harness.

As for *Back to the Future*, the movie's representation of time travel is flawed. It fails to reconcile the grandfather paradox: if you went back in time and killed your grandfather, would you cease to exist? On one hand, if your grandfather was dead, there's no way you could be born; on the other hand, if you weren't born, you couldn't grow up to travel back in time and kill your grandfather—so you would be born anyway!

J. Richard Gott, an expert on the physics of time travel, issues a clear rebuke to the skeptical question Hawking has posed to time travel enthusiasts: "Why haven't we been overrun by tourists from the future?" The short answer: because time travel will only be possible within the period that a time machine is in existence. If you're interested in the technical explanation of these ideas, a good place to start is Gott's book, *Time Travel in Einstein's Universe*, which argues that both wormhole time travel and cosmic string time travel (a complicated method that relies on string theory) forbid time travel into eras in which a time machine hasn't been built. So if you haven't met your future self yet, don't worry. If you hang around long enough, you just might meet many more of you.

○ ○ ○

Is string theory a breakthrough in cosmology or the product of a bunch of theoretical physicists spending too much time indoors?

TOO TANGLED FOR MOST OF US TO UNRAVEL

String theory is an extremely difficult new way of thinking about the universe. On a basic level, it says that fundamental particles resemble strings, not points, in space. Particles are always moving in a single direction until they interact with other particles that change their direction. Each particle's string is the trace of that particle's path through space. The string may be wavy, bent, or even looped. When we observe a particle moving in some line or wave, what we're actually seeing is the string vibrating. We're watching the vibration move from one end of the string to the other.

When another particle interacts with this first particle in some way, their strings join up, creating a longer string. The diagrams in Hawking's book show this rather well (figures 11.3–11.4). What the book doesn't show well is how this new description of particles leads to a totally new conception of the dimensions of space-time. For instance, string theory successfully predicts the actions of particles only if we assume that our universe has eleven dimensions instead of four. How does the way we imagine particles as strings lead to this dramatic, bizarre conclusion?

It doesn't, at least not intuitively. Like several strains of logic discussed in *A Brief History of Time*, the answer lies in the sophisticated math used in string theory. These extra dimensions are just as arcane: if they indeed exist, they're each curled up into a tiny sphere with the circumference of 10^{-33} centimeters. True, there might be whole separate universes with these dimensions, but like the whole other universes of antiparticles and antipeoples, we'll never meet them.

Although the premises of string theory aren't too difficult to understand, the implications of the theory are still unknown. String theory allows for new solutions to questions about the origin of the universe and about time travel, but these are still being developed. String theory has helped scientists to understand the ways forces work, but it hasn't yet deeply changed our conception of space and time. Moreover, there isn't just one string theory, but four! They seem to be related to one another, but no great unifying trick has been discovered. (For a comprehensive study of string theory written for the general public, check out the two-volume *Superstring Theory* by Ed Witten et al.). At this point, the string theories are a promising, curious, and perhaps necessary transition phase on the way to a new physics of the future—and to a theory of everything.

Where does Hawking see cosmology going? Can we realistically ever reach an ultimate theory of the universe?

IT'S OUT THERE SOMEWHERE

Hawking is positive about the future of cosmology and of science in general, though he's cautious as well. To put it another way, he's very confident in the beauty of science but less confident in the ability of humans. Although the title of Hawking's newest book, *The Theory of Everything,* made a big splash when it was announced, it's a red herring. Hawking hasn't really come any closer to pinning down such a momentous theory. He's only come a little closer to being able to explain current theories. In essence, this new title is a repackaging of the material he previously introduced most successfully in *A Brief History of Time.*

But what about the possibility of ever discovering a theory of everything? Hawking believes it's possible. Hawking breezily dismisses the idea that even if we did formulate such a theory, we could never know it to be true, since its accuracy could be tested only by the future. (That is, the theory could make accurate predictions for the present, but the future might always have a surprise in store for us.) Hawking is right to call this kind of thinking useless. After all, the same thing could be said about any scientific theory.

Hawking believes that a theory of everything exists because he believes in a grand universal order. His research has proved to him that

H.A.L. and his pals

Hawking has warned that in the future, mankind may be taken over by robots. That's right—robots. In a 2001 interview with the German magazine ***Focus,*** Hawking said, "In contrast with our intellect, computers double their performance every eighteen months. So the danger is real that they could develop intelligence and take over the world." At moments like this it's some comfort that Hawking is a theoretical astrophysicist, not a computer scientist or prophet.

beneath every seemingly inexplicable natural event is a natural law governing it. What's more, these laws of nature can be expressed by mathematical equations and scientific theories. Although Hawking's faith in an omniscient and omnipotent god dissipates as early as his discussion of the no-boundary universe, the last few pages of *A Brief History of Time* show a religious-like faith in science. In opposition to mankind, which is mortal and self-destructive, Hawking portrays a nature that is elegant, perfect, and all-knowing. By the time the book winds down to its conclusion, Hawking has transformed his science into his own form of theology. He worships the pure scientific perfection of the universe like a monk contemplating Scripture.

Ironically, then, at the point in the book where Hawking seems to have most abandoned religion, he replaces it with a structure of similar majesty and austerity. For Hawking, the universe has become a deity that inspires a kind of religious devotion. The God that Hawking writes about in the final pages of his book isn't the same one he refers to in the opening pages, in which he uses physics to launch inquiries into the existence of the God of the Judeo-Christian tradition. Instead, in these final pages, we realize that Hawking has erected a new god, a god whose name has come to represent the order of the universe in all its precision and beauty.

Despite Hawking's optimism about one day discovering a theory of everything, we can't help but feel that mankind's search for this theory will ultimately end in failure. The more we learn about the universe, the more it astonishes us. Our own efforts to grasp it only seem to uncover deeper mysteries. In turn, our determination to force this shifty universe into a single theory seems to be increasingly hopeless. For how could a God as perfect as the one who created this vast and complicated universe ever be caught, red-handed, in the midst of the act of creation?

A WRITER'S LIFE

A Rough Road

Against all odds, Hawking has coped with the burden of ALS and become perhaps the foremost physicist of our time.

○ ○ ○

ILLNESS

Many people don't realize that Stephen Hawking wasn't born with ALS (also known as Lou Gehrig's disease) but developed symptoms only in his early twenties. At that time, in 1963, medical science, much like cosmological science, couldn't compare to what it is now. Not surprisingly, as Hawking tells it, his experience in hospitals was nightmarish.

Though Hawking wasn't especially athletic or graceful before the disease struck, he did participate in some sports and even managed to join an intramural rowing team as a coxswain in his first years at Oxford University. During his third year there, however, he noticed that his coordination was deteriorating, and he found himself unable to manage the boat successfully. Soon, he found himself bumping into things and falling down for seemingly no reason.

When basic tests offered no insight, Hawking was taken into the hospital, where the real horrors began. He tells this part of his story in a somewhat grim tone, with open disdain for the so-called experts who reviewed his case. On his web page, Hawking describes the wide variety of tests performed on him: "They took a muscle sample from my arm, stuck electrodes into me, and injected some radio opaque fluid into my spine, and watched it going up and down with x-rays, as they tilted the bed. After all that, they didn't tell me what I had, except that it was not multiple sclerosis, and that I was an atypical case." He remained in the hospital for two weeks and left feeling depressed by the news that his condition, whatever it was, would likely only deteriorate.

Later, doctors diagnosed Hawking with ALS and predicted he would die in only a year or two. The doctors had no reason to think otherwise: in its most common form, ALS kills the vast majority of its victims (including Lou Gehrig) within this time frame. The fact that Hawking has survived and gone on to such miraculous achievements only adds to his stature in the popular imagination. But as we might expect, Hawking has a more scientific response: "ALS seems to be a condition that can result from different causes. The variety I have must be different from the most common form, which kills in two or three years. Maybe my ALS is [a rare kind] caused by bad absorption of vitamins. My wife says I'm an alien in the morning before I have my vitamins."

Despite the incredible fact of his survival—which Hawking calls the greatest achievement of his life—the disease makes his life more difficult than the public might perceive. After all, relatively soon after being diagnosed with ALS, Hawking realized that "the rest of the world won't want to know if you are bitter or angry. You have to be positive if you are to get much sympathy or help." His daily routine is extremely ordered, laborious, and, in general, fantastically slow. Hawking can't eat, wash, or dress himself without complete assistance from his full-time nursing staff. With

Starting over

After Hawking lost his speech ability, he was forced to teach himself an entirely new method of thinking about physics. Normal theoretical work requires a great deal of math and therefore the writing and solving of countless equations. Hawking was no longer suited for this task. Kip Thorne, one of the most brilliant minds in theoretical physics today, described Hawking's transformation: "As Stephen gradually lost the use of his hands, he had to start developing geometrical arguments that he could do pictorially in his head. He developed a very powerful set of tools that nobody else really had." Hawking's memory, already remarkable, became even more acute. His new emphasis on pictorial representation inspired him to revise ***A Brief History of Time*** and later to write ***The Illustrated Brief History of Time.***

the help of the internet, he can now at least do research and access scientific papers and correspondence relatively quickly. But writing, like talking, is quite painstaking.

Hawking has been unable to use his vocal chords since 1983, when he required an emergency, lifesaving tracheotomy due to complications from pneumonia. He now communicates through a machine called the Equalizer, designed by Walt Woltosz, a California programmer. The Equalizer, which is built into Hawking's wheelchair, has a cursor controlled by a switch that Hawking holds in his hand. By pressing and releasing the switch, Hawking is able to navigate his cursor through the computer's 3,000-word vocabulary and select entries. If a word he is searching for isn't in this word library, he must use another screen that allows him to spell it out. The computer then speaks the word in an amplified voice that mimics human intonation, though with an American accent—a source of constant frustration for Hawking. After fifteen years or so of practice, Hawking can now articulate about ten to fifteen words per minute. A forty-minute speech, therefore, can take him up to forty hours to write.

But these technologies have been feasible only because Hawking's disease, which atrophies the nerves controlling muscle movement, hasn't yet done significant damage to the muscles in his hand. Sadly, since ALS is degenerative, Hawking may one day lose control of his hands as well. If this were to happen, Hawking would still have a fully functioning mind but would not be able to communicate or express his thoughts in any way. It's a haunting possibility, one that, like many other details of Hawking's life, befits science fiction. But Hawking remains optimistic. "While I am alive," he once told Larry King, "I will make sure I communicate one way or another."

SCIENCE AND GENIUS

Oddly enough, Hawking's genius didn't emerge until some time after he was diagnosed with ALS at the age of twenty-one. He was a bright child, but not one who appeared destined for greatness. He grew up in Highgate, an area outside Oxford, and attended the local schools there before moving to St. Albans, north of London. The first thing named after him was not "Hawking radiation," a term sometimes used for the radiation that is emitted by black holes—it was "Hawkingese," the term his classmates used to describe the language spoken by the awkward, rakish young Hawking, who had a dramatic lisp. Outside of school, Hawking had the unique habit of inventing extraordinarily complicated board games that took days to complete. Hawking wasn't as unlikely a candidate for greatness as Einstein, whose poor reputation as a student is legendary, but Hawking wasn't exceptional either.

Hawking was awarded a scholarship to study physics at Oxford University, where things started to turn for him. Despite being a rather inattentive student—he worked the bare minimum of hours necessary to fulfill his requirements—Hawking received first-class honors in physics. But only barely. Even though he aced the theoretical sections of his exams, he had to have a private makeup interview in order to pass the technical sections. By the time Hawking's illness had become debilitating, he had entered Cambridge University as a doctoral candidate in physics and had begun to study black holes with Roger Penrose. This work was what led to Hawking's first famous discovery, the Hawking radiation emitted by black holes.

Now a father of three, Hawking credits his wife, Jane, with much of his success at overcoming his disease and living a highly accomplished life. Jane even makes an appearance in the text of *A Brief History of Time,* as the person who inspired Hawking to keep pursuing cosmology in those first years after being diagnosed with ALS. But the inspiration Jane provided wasn't entirely romantic: if Hawking wanted her to stay with him, he had to have some money, and the only way to do that would be to finish his Ph.D. So as sick as he was, he finished that Ph.D. And she married him.

ROLE IN THE SCIENTIFIC COMMUNITY

Hawking is one of a small group of celebrities in various fields whose fame dwarfs that of their peers. The figures who accomplish such a feat often do so for reasons other than the sheer magnitude of their abilities within that profession. Anna Kournikova, for instance, may be the most popular player currently on the women's tennis tour, even though she has never won a single tournament. We might argue that Hawking's international fame is because of the monumental sales of his books and because, as he puts it, "I fill the role model of a disabled genius." (Though he calls the title of genius "rubbish, just media hype.") Guest appearances on *Star Trek* and *The Simpsons*, along with a starring role in the film version of *A Brief History of Time*, haven't hurt his fame either. Hawking is a contemporary icon, a man famous for being great at a discipline that few understand. As fellow physicist John Gribbin has written, Hawking "has reached the heady status of being famous for being famous."

Hawking is a contemporary icon, a man famous for being great at a discipline that few understand.

But, as Gribbin admits, Hawking hasn't earned fame without good reason. After all, Hawking is responsible for discovering the way we now believe the universe began—with a big bang singularity. This idea is perhaps the biggest reason for his popularity. But on top of this, Hawking's proposal that black holes emit radiation earned him great respect from his peers. Although it was a relatively esoteric finding, Hawking's technique was revolutionary. It was the first discovery to combine the two of the great physics theories of the twentieth century—the uncertainty principle and Einstein's theory of general relativity. Since it's now popularly believed that a theory of everything will reconcile these two theories, Hawking's work on black holes seems to provide a first step toward this proposed grand theory.

Nevertheless, Hawking still must bear the grudge of his colleagues for his incredible fame. One fellow scientist has said that Hawking is "esteemed but not revered" by the majority of physicists. Gribbin writes, "Whatever anyone may tell you, [Hawking] is not the greatest scientific thinker since Albert Einstein: that honor probably belongs to the late Richard Feynman" (one of the fathers of quantum electrodynamics, the theory of virtual particles). Still, for every physicist furiously typing away on his university web site, there are hundreds of figures in the popular press making the now-familiar parallels between Hawking and Einstein, concluding that Hawking is the most intelligent man alive. Read the book and decide for yourself: how would we think about the universe now if it weren't for Hawking's scientific—and writerly—contributions?

A POP-CULTURE ICON

Although *A Brief History of Time* gave Hawking the distinction of being one of the best-selling authors of all time, his celebrity is even more astonishing. Certainly, it's impressive that a theoretical astrophysicist should become a celebrity at all. Only Einstein, among physicists, shares such high renown. But celebrity doesn't even accurately describe Hawking's status in popular culture. Hawking is more than just celebrated—he is an icon.

Hawking's fame is so great that he has become instantly recognizable. Jennifer Aniston, for instance, is a celebrity: most Americans know something about her or would be able to recognize her on the street. But icons become more than just familiar faces—they enter the popular consciousness in a way normal celebrities cannot. Just as we reduce Marilyn Monroe to the platinum hair, the sexy pout, and the billowing skirt, we conceive of Hawking as the robotic voice and the slumped figure in the high-tech wheelchair. These images, or associations, are deeply embedded in the public consciousness and persist long after a generation's other celebrities have faded away. Werner Heisenberg was something of a celebrity, but we have nothing to associate with him. The general public has forgotten him. But we still have Einstein's glasses, his white shock of hair, and his shaky German accent.

Thanks to *A Brief History of Time*, we think of Hawking in these iconic terms. His fame even allowed him the opportunity to go head to head with Einstein and Newton—in a game of poker. No, he didn't slip through a wormhole to challenge these two old masters; he just made a cameo on the 1993 season finale of *Star Trek: The Next Generation*. And he got to give Einstein (albeit an actor playing Einstein) a now famous retort:

> **HAWKING:** I raise $50.
>
> **LIEUTENANT COMMANDER DATA:** All the quantum fluctuations in the universe will not change the cards in your hand. I call.

NEWTON: You are bluffing.
EINSTEIN: And you will lose.
HAWKING: Wrong again, Albert. *[Hawking wins the hand with four of a kind]*.

An even more famous appearance came on a 1999 episode of *The Simpsons*, when an animated Hawking floats down to Springfield in a flying wheelchair and saves Lisa and her fellow Mensa members from Homer and a raucous mob:

HAWKING: I don't know which is the bigger disappointment, my failure to formulate a unified field theory or you.
HOMER: I don't like your tone. . . . *[To Lisa]* Did you have fun with your robot buddy?
LISA: Dad! Oh, Dr. Hawking, we had such a beautiful dream. What went wrong?
HAWKING: Don't feel bad, Lisa. Sometimes the smartest of us can be the most childish.
LISA: Even you?
HAWKING: No, not me. Never.

Hawking has also appeared as a guest on *Larry King Live*, been named one of *People* magazine's most intriguing people, shown up in advertisements for British Telecom and U.S. Robotics, and even supplied vocals to a Pink Floyd song. He has fan clubs and groupies, and has been referenced in such sitcoms as *Murphy Brown* and *Seinfeld.* All this doesn't even include the stories about Hawking that have been featured on television shows and in publications too numerous to count. Major newspapers like *The New York Times*, the *Washington Post*, and the *Wall Street Journal* have not only reviewed Hawking books and run feature pieces on him, but they have even published articles dedicated solely to the phenomenon of his fame.

A *Brief History of Time* alone has created a media giant. It spawned a six-part television film, *Stephen Hawking's Universe,* featuring Hawking and airing on PBS, where Carl Sagan's *Cosmos* also aired; an interactive CD-ROM; *The Illustrated Brief History of Time*; and *Stephen Hawking's A Brief History of Time: A Reader's Companion.* In 1992, Hawking was the subject of a documentary by Errol Morris, one of the most famous and accomplished documentarian alive. Although it's named after Hawking's book, Morris's film focuses primarily on the story of Hawking's life, featuring interviews with family members, friends, and colleagues. Hawking is also the subject of no less than eleven biographies, with many more planned.

A BRIEF HISTORY

Mass-Market Science

Hawking's book was a publishing phenomenon, making complex science accessible to a wide audience like never before.

○ ○ ○

WE MIGHT EXPECT THE SIGNIFICANCE of *A Brief History of Time* to lie in its scientific achievement—after all, Hawking is renowned for his revolutionary scientific breakthroughs, and the book is an explanation of his life's work. But in fact, the opposite is true. As a scientific document, the book is rather insignificant. Most of Hawking's important discoveries were made years before the book's publication. His work on singularities was conducted in the late 1960s and early 1970s. His significant findings on black holes were announced in the early 1970s. By the time *A Brief History of Time* was published, these phenomena were as familiar to his contemporaries as Newton's findings on gravity. *A Brief History of Time* was published about fifteen years after Hawking's last major cosmological finding (and fifteen more years have passed since).

So, unlike many books that we might have found in the science section of a bookstore in 1988, *A Brief History of Time* didn't claim to provide new information. By the standards of the field of cosmology, a fifteen-year-old discovery is relatively dated. Indeed, there were many other significant discoveries in the years just preceding the publication of the book. Perhaps most famous of these was the articulation of string theory, to which Hawking gives some attention.

What makes *A Brief History of Time* remarkable isn't its scientific ingenuity but the fact that it's targeted at the general reader. It's a bit misleading to way the book was sold in the science section of a bookstore when it appeared in 1988—it wasn't. A polite salesperson would have pointed you back to the front of the store, the middle of the window display. *A Brief History of Time* was the first book of recent years to explain successfully (for the most part) the workings of the universe—and the first therefore to sell on such an enormous scale. Although other popular cosmology books had already been published, such as Carl Sagan's *Cosmos*, Hawking's was the first to make some of the field's most complicated theories understandable to non-experts. His subject was not simply the Earth and the stars, but the nature of time, space, and existence—subjects that previous physics writers hadn't approached outside of science journals and academic newsletters.

In short, the legacy of *A Brief History of Time* is more cultural than scientific. The sheer popularity of the book forced booksellers and publishing houses to take the genre of popular science more seriously. Although science bestsellers such as Jared Diamond's *Guns, Germs, and Steel* and Matt Ridley's *Genome* are now relatively common, this wasn't always the case. Not until *A Brief History of Time* did publishers fully appreciate how well a science book could sell. Not surprisingly, they began to put more money into promoting science-themed projects. Perhaps even more impressive, Hawking's book all but created a new popular genre within science books—the trade physics book. Now, books about everything from relativity to quantum physics are released widely and with much fanfare, coming out in neatly designed paperback editions a year later.

It's not enough simply to credit Hawking's lucid prose for the popularity of his book. After all, critics at the time noted that despite Hawking's clear writing, many of the book's concepts remained beyond the grasp of the average reader. *Boston Globe* columnist David Warsh wrote, "A *Short*

All fun and games

In public interviews, Hawking usually plays down his guest appearances on sitcoms like ***Star Trek*** and ***The Simpsons,*** saying they were great fun but "not to be taken seriously." However, he once admitted that for the millennium new year he was planning a "***Simpsons*** fancy dress party." He went on to say, in classic Hawking form, "People are coming as Springfield characters. The great thing is I can go as myself."

History of Time is a seriously difficult book"—so difficult, it seems, that Warsh was unable even to grasp the book's title. "It's not altogether impossible to follow, at least at some level. But it is difficult to believe that even a substantial fraction of those who bought it have even read it, much less understood it. . . . What's going on?" In *The New York Times,* Michiko Kakutani wrote, "[I]t is hard for the lay reader to grasp all of Mr. Hawking's arguments involving [his] new vision of the universe." Kakutani referenced the "dense passages dealing with 'imaginary time,' 'string theories' and 'inflationary' models of the universe, which this reader found impossible to follow." The notion of imaginary time especially frustrated Martin Gardner, who reviewed Hawking's book for the *New York Review of Books.* "It is hopeless to explain the new model in any detail because it makes use of . . . 'imaginary time.' . . . Hawking makes no attempt to explain his model except by a vague analogy."

So, to repeat Warsh's question, what's going on? What is it about these abstract theories that have driven a generation of readers to Hawking? It would be difficult to argue that the book's success depended on the time in which it was published, since the book continues to sell well today—well enough to spur an updated, tenth-anniversary edition of the book several years ago.

Even though we can't attribute the popularity of *A Brief History of Time* to the historical moment in which it was published, there's something to be said for its larger historical context. The last thirty years—roughly the span of Hawking's career—has seen an unprecedented faith in science among the general population. This new acceptance of the power of science to answer big questions has taken place only in recent decades, as science has begun to catch up with these questions. Take, for instance, the question of how the universe began. Scientists may have always had theories that attempted to answer this question, but only in the past fifty years has physics been able to explain how such a beginning might have come about. And only in the last ten years have scientists become able to point to real evidence of the big bang theory. It's one thing to know that such evidence exists, but to see the evidence for ourselves, it became necessary to read Hawking's book.

The second half of the twentieth century has even seen the world's highest religious authorities accept theories that were once considered heretical or baseless. Even the Catholic church, as Hawking notes, pronounced the big bang theory to be in accordance with church doctrine as far back as 1951. Although the pope might not have admitted it then, his statement reflected a growing convergence of religion and science in the effort to answer great questions of existence. When searching for the answers to our biggest questions, science is at least as capable of answering them as religion, if not more so. As the first bible of modern-day cosmology, then, it's no wonder that *A Brief History of Time* has sales figures almost on par with the Bible.

The essayist Charles Krauthammer attributed the success of Hawking's book to its similarity to the Bible. He wrote that even if people don't completely understand either book, the simple possession of them "may express a certain reverence for learning." When asked what he thought of the fact that many people who own Hawking's book likely don't finish it, Krauthammer said, "Not many people read their Bibles, either. But they like having them around." Indeed, the difficult parts of Hawking's book may only add to his mythic status. What religious text lacks moments of confusion, complication, or frustration?

For such books to remain alive in the public consciousness, they must perform two functions: they must help us to answer our biggest questions, but they must also pose new ones as well. Hawking tells us how the universe began, but in doing so, he questions the idea of a "beginning" in the first place: in his "no boundary" model, the universe has no conventional beginning or end. It simply always exists. Similarly, although his explanations help to demystify the workings of wormholes and black holes, Hawking goes on to ask if maybe our entire universe is itself a black hole, separated by a wormhole to another universe. Until a book comes along that's able to answer these new questions, it's likely that this bible will have no new testament.

EXPLORE

Other Books of Interest

A *Brief History of Time* is just the first of a number of books that have tried to put complex science in plain English.

○ ○ ○

BY STEPHEN HAWKING

THE UNIVERSE IN A NUTSHELL

(Bantam, 2001)

Essentially a recap of *A Brief History of Time*, this book aims to be more easily accessible to the lay reader, employing greater use of diagrams. *The Universe in a Nutshell* hasn't sold nearly as well as its predecessor, perhaps because it introduces theoretical concepts that many readers and critics actually have found *more* difficult to follow.

THE FUTURE OF SPACETIME

(Norton, 2002)

Hawking joins Kip S. Thorne, Igor Novikov, Timothy Ferris, and Alan Lightman in a collection of essays on the issue of space-time. Each contributor writes from the perspective of his own specialty. The collection is unique in its multidisciplinary approach, combining essays by writers from different backgrounds. Lightman, the author of the bestselling novel *Einstein's Dreams*, discusses the relationship between writing science-based fiction and studying theoretical physics. Thorne imagines the future's wedding of the theories of general relativity and quantum mechanics. Hawking argues why he thinks time travel is impossible, tacitly rebuking Gott's claims (see below). *The Future of Spacetime* is a nice introduction to the writing of today's most important scientific minds.

ABOUT HAWKING

STEPHEN HAWKING: A LIFE IN SCIENCE

by Michael White and John Gribbin (Dutton, 1992)

Although it's somewhat dated, this is the most reputable of the Hawking biographies. It succeeds because both authors are excellent writers and have excellent scientific minds (White is the director of Science Studies at d'Overbroeck's College, Oxford, and Gribbin is a highly respected scientist and author of *Cosmic Coincidences*, among other popular books). The book does an excellent job of explaining Hawking's science to the lay reader while providing a cohesive, dramatic narrative of Hawking's life. The biography is mostly supportive of Hawking and his theories, but since both authors work within the field of astrophysics, they have access to some sordid details about Hawking's career. One example is a story about how Hawking nearly managed to ruin the career of a colleague with whom he quarreled. All in all, this is the closest thing to a definitive Hawking biography to date.

MUSIC TO MOVE THE STARS: A LIFE WITH STEPHEN HAWKING

by Jane Hawking (Trans-Atlantic Publications, 1999)

Hawking's former wife tells the story of her struggle to raise three children and support an ailing husband. The description of Hawking leaving his dutiful, saintlike wife for one of his nurses—the wife of the man who designed his speech synthesizer—doesn't exactly make for a sympathetic portrayal of the scientist's personal life. Jane describes Hawking as a cruel domestic dictator, an "all-powerful emperor," and a "masterly puppeteer." To her credit, Jane doesn't paint the most glowing self-portrait either, describing the fading of their love, that she found it "difficult—unnatural, even—to feel desire for someone with the body of a Holocaust victim and the undeniable needs of an infant." With her husband's consent, Jane began an affair of her own with a family friend. For those interested in the less glamorous aspects of a genius's life, read about all this and more in Mrs. Hawking's 600-page tell-all biography.

ON RELATED SUBJECTS

QUANTUM LEGACY

by Barry Parker (Prometheus Books, 2002)

For those who enjoy Hawking's digressions into the lives of important scientists, *Quantum Legacy* is an easy, friendly read. An energetic narrator, Parker tells the story of Einstein's discovery of quantum mechanics as if it were a whodunit, with Einstein playing the role of Sherlock Holmes. Parker also pays close attention to the various Watsons, delving into the lives of figures like Max Planck, Niels Bohr, Werner Heisenberg, Erwin Schrödinger, and others who contributed to quantum theory. By explaining each aspect of the theory in the order in which it was originally discovered, Parker breaks quantum mechanics down to manageable units and lets readers share in the excitement of the physicists' discoveries. Parker also extends the narrative into the present, exploring the practical applications of quantum theory. *Quantum Legacy* is a touchingly human approach to what is often regarded as one of the coldest and most abstract scientific theories.

TIME TRAVEL IN EINSTEIN'S UNIVERSE

by J. Richard Gott (Houghton Mifflin, 2001)

Gott, a professor of astrophysics at Princeton University, is so renowned for his expertise on time travel that not only do "the neighborhood children think I have a time machine in my garage," but "even my colleagues sometimes behave as if I have one." Gott is one of the foremost cosmologists of our time and also a great storyteller, with a child's fascination for time travel and all the popular lore surrounding it (the book even begins with a discussion of H.G. Wells's *The Time Machine*). Despite this childlike wonder, Gott's book is a rigorous discussion of the physics of time travel, built directly on Hawking's own theories. This discussion leads to a debate about different possible models of the universe—which, in turn, has been known to lead to migraines. But for those who are fascinated by the idea of time travel and unsatisfied with Hawking's brief assessment, *Time Travel in Einstein's Universe* is as clear and comprehensive a book as you can find on the topic. Gott is a patient teacher, trying his best to convey the science to the lay reader. If nothing else, look for his enthusiastic assertion that time travel is not only possible, but has happened already!

THE BOOK OF NOTHING

by John D. Barrow (Vintage, 2000)

The popular Barrow is a professor of theoretical physics at Cambridge University and the author of such other titles as *Theories of Everything, Impossibility*, and *Pi in the Sky*. Barrow is capable of explaining science clearly, but his approach is more fluid than most authors in the science genre. For Barrow, science is the way man makes sense of the world around him, so Barrow draws upon myth, pop culture, and literature as readily as he draws upon physics. Einstein helped us to make sense of our world, but so did Shakespeare and Freddie Mercury.

In his provocative book, Barrow examines the notion of nothing, exploring its various manifestations in the universe. He covers topics such as black holes, the beginning and end of the universe, vacuums, and the history of the number zero in different traditions. Barrow fleshes out many topics Hawking introduces, taking them in unforeseen directions. *The Book of Nothing* may not be the book for those looking for hard facts and explanation. But on its own terms, the book leads the willing reader on a sometimes haunting, sometimes rollicking adventure through both the universe and the human imagination.